CGP

GCSE

Core Science

Answer Book

Practice Exam Papers

Higher Tier

These practice papers won't make you better at science

... but they will show you what you **can** do, and what you **can't** do.

These are GCSE papers, just like you'll get in your exams — so they'll tell you what you need to **work at** if you want to do **better** on the day.

Do an exam, **mark it** and look at what you **got wrong**.
That's the stuff you need to learn.

Go away, **learn** those tricky bits, then **do the same exam again**. If you're **still** getting questions wrong, you'll have to do even **more practice** and **keep testing** yourself until you keep getting **all** the questions right.

It doesn't sound like a lot of **fun**, but it **really will help**.

The three big ways to improve your score

1) **Answer all these exams**
These practice papers contain all the types of question that have come up year after year in GCSE exams. If you can do all these, you should be able to do all the questions in your exams.

2) **Keep practising the things you get wrong**
The whole point of a practice exam is to find out what you don't know*. So every time you get a question wrong, revise that subject then have another crack at it.

* Use the mark scheme in this booklet to help you see where you dropped your marks.

3) **Don't throw away easy marks**
Always answer the question the way it's asked — if it's worth three marks, give three points in your answer. Always double-check your answer and don't make silly mistakes — obvious really.

Contributors: Mike Dagless, Mark A Edwards, James Foster, Rebecca Harvey, Judith Hayes, Frederick Langridge, Adrian Schmit, Claire Stebbing, Pat Szczesniak, Sophie Watkins, Chris Workman.

Working out your Grade

- Do each of the six papers in this pack.
- Use the answers and mark scheme in this booklet to mark them.
- For each paper, divide your score by the total number of marks available for that paper. (Each paper 1 has a total of 32 marks, and each paper 2 has a total of 30 marks.) Multiply this value by 100 to get a percentage.
- Find your average percentage for the whole exam (all six papers).
- Look it up in this table to see what grade you got. If you're borderline, don't push yourself up a grade — the real examiners won't.

Average %	85+	74 - 84	61 - 73	47 - 60	37 - 46	30 - 36	Under 30
Grade	**A***	**A**	**B**	**C**	**D**	**E**	**U**

Stick your marks in here so you can see how you're doing

		Physics	Chemistry	Biology	Average %	Grade
Paper 1	First go					
	Second go					
	Third go					
Paper 2	First go					
	Second go					
	Third go					

Important!

Any grade you get on one of these practice papers is **no guarantee** of getting that in the real exam — **but** it's a pretty good guide.

Biology Paper 1

Answer			Explanation and tips
1.	**1A**	2	Patient A fits the bill.
	1B	1	People with diabetes can't control their blood sugar level very well so it's likely to be high.
	1C	4	The heaviest smoker is most likely to get lung cancer.
	1D	3	High blood pressure, smoking and obesity all increase the chance of getting heart disease.
2.	**2A**	3	Plants remove carbon dioxide from the air for photosynthesis.
	2B	2	Cattle release methane gas, which enters the atmosphere.
	2C	4	Excess fertiliser can be washed into rivers and streams, causing pollution.
	2D	1	Carbon dioxide is produced when petrol or diesel is burnt.
3.	**3A**	3	
	3B	2	The question asks why it wasn't accepted by the scientific community not the religious community.
	3C	2	
	3D	1	
4.	**4A**	1	Genes control a person's characteristics.
	4B	3	
	4C	2	
	4D	3	Carlo and Louis have an identical gene pool to pass on, so their children will share similar characteristics.
5.	**5A**	1	It's the one that causes the biggest clear patch around the disc (where the antibiotic has killed the bacteria).
	5B	4	The dependent variable is the thing that you measure.
	5C	3	
	5D	2	If the bacteria have developed resistance the antibiotic wouldn't kill them.
6.	**6A**	4	Hormones always travel in the blood.
	6B	1	IVF doesn't guarantee having twins and you can't choose what colour eyes the baby will have.
	6C	2	
	6D	3	The introduction states that they can cause abdominal pain and dehydration.
7.	**7A**	3	300 - 50 = 275
	7B	1	
	7C	3	Be careful — the other three options may be likely, but you can't tell from the graph.
	7D	2	Look for the point on the graph where the lines cross.
8.	**8A**	2	A stimulus is a change in environment, e.g. someone poking your neck.
	8B	2	Reflexes are automatic, if you have to think about what response to give then it's not a reflex action.
	8C	3	The more you repeat an experiment the more reliable your result becomes.
	8D	2	

Biology Paper 2

1 **a)** **i)** *Any one of,* e.g. nicotine / ecstasy / caffeine *[1 mark].*

ii) Class A *[1 mark]*

b) The liver can become damaged as it removes toxic alcohol. These cells may die, forming scar tissue that stops the blood reaching the liver *[1 mark].*

If the liver can't do its normal job of cleaning the blood dangerous substances start to build up and can damage the rest of the body.

c) *Any one of,* tar damages the cilia, preventing them from moving bacteria and mucus out of the lungs / tar contains carcinogens, which could cause cancer *[1 mark for effect, 1 mark for reason].*

---

2 **a)** X *[1 mark].* The type of organism found at the site is often found in polluted water / mayfly nymph are found in area Y so it's likely to be less polluted *[1 mark].*

b) By the waste from the industrial site *[1 mark].*

c) *Any one of,* e.g. they could contain useful products which would be lost with the species / loss of species from an ecosystem could unbalance it *[1 mark].*

d) mayfly nymph *[1 mark]*

e) *Any one of,* e.g. the population has increased / standards of living have increased *[1 mark].*

The human population has increased enormously over the last 200 years. This means the amount that we are polluting the Earth has also risen.

f) E.g. to make sure the needs of today's population are met without harming the ability of future generations to meet their own needs *[1 mark].*

---

3 **a)** The harmless pathogen carries antigens which stimulate antibody production *[1 mark].* If you become exposed to the pathogen again your body can rapidly produce antibodies against it so you won't get ill *[1 mark].*

b) E.g. it reduces the spread of disease *[1 mark].*

c) E.g. some people may suffer from side effects / some people think that immunisation can cause other disorders *[1 mark].*

---

4 **a)** pathogens *[1 mark]*

b) *Any one of,* e.g. bacteria / viruses / fungi *[1 mark].*

c) E.g. skin forms a barrier, blood clots seal wounds *[2 marks].*

d) They attach to antigens on the surface of the pathogen and brings about its death *[1 mark].*

---

5 **a)** **i)** *Any two from,* e.g. large/brightly coloured petals / scented / produce sticky pollen / produce nectar *[1 mark for each]*.

ii) *Any one of,* e.g. minerals in the soil / water / light *[1 mark]*.

b) **i)** They fix nitrogen, providing the plant with essential nitrates *[1 mark]*.

In return the plant provides the bacteria with a constant supply of sugar.

ii) Mutualistic *[1 mark]*.

---

6 **a)** **i)** *Any one of,* e.g. levels of fishing increased / larger/better fishing nets were used so more cod were caught / increase in the amount of harmful pollution *[1 mark]*.

ii) E.g. introduction of fishing quotas / restrictions on the type/size of nets that could be used *[1 mark]*.

b) *Any two from,* e.g. setting up protected habitats / introducing new legislation to protect them / captive breeding / creating artificial ecosystems / education programmes *[1 mark for each]*.

Chemistry Paper 1

Answer			Explanation
1.	**1A**	2	Carbon can be used to extract metals that are below it in the reactivity series.
	1B	4	
	1C	1	This is the only metal in the list.
	1D	3	
2.	**2A**	3	
	2B	2	
	2C	1	
	2D	4	
3.	**3A**	3	3 and 4 seem the most likely, but fitting together like a jigsaw is evidence, so it must be 3.
	3B	1	
	3C	4	We now know that the continents move by convection currents, not by ploughing through the ocean beds.
	3D	2	Radioactive decay causes convection currents, which cause the plates to move, so it can't be either of those.
4.	**4A**	1	The less reactive an element is, the less likely it is to combine with other elements and form compounds.
	4B	2	
	4C	3	
	4D	3	It has to be one or three but one does not balance.
5.	**5A**	3	Find 25 on the x axis and go up until you hit the line of best fit, then read off the answer from the y axis.
	5B	3	A thermometer that measures up to 350 °C is needed, the one that measures up to 600 °C is less accurate.
	5C	2	Add the numbers together and divide the total by three to find the average.
	5D	1	Helen's results differ from those in the graph, so another factor may have affected the results.
6.	**6A**	3	An alloy is a mixture not a compound.
	6B	2	
	6C	4	
	6D	1	Nitinol can be bent and twisted but on heating it goes back to its original 'remembered' shape.
7.	**7A**	1	All you have to do is count up the number of carbons and hydrogens in hydrocarbon Z.
	7B	3	Remember, alkenes have double carbon bonds so it must be X and Y.
	7C	2	Alkanes can't be turned into polymers, whereas alkenes can.
	7D	4	Bromine water is a test for alkenes (it turns from brown/orange to colourless).
8.	**8A**	3	It is the only fuel which does not produce carbon dioxide.
	8B	2	$32\,500 \div 1625 = 20$ litres
	8C	1	
	8D	2	Petrol and diesel are mixtures of compounds and hydrogen is an element.

Chemistry Paper 2

1 a) The aluminium oxide acts as a catalyst *[1 mark]*.

b) *Any one of,* e.g. alkenes are formed, which are useful for making polymers / the process helps to match the supply of certain fractions with the demand for others. *[1 mark]*

c) Gas A: propane *[1 mark]*. Gas B: ethene *[1 mark]*.

Gas A has no double bonds so it must be an alkane.

---

2 a) Volcanic eruptions from inside the Earth's crust *[1 mark]*.

b) Green plants evolved and photosynthesised, producing oxygen *[1 mark]*.

c) Nitrogen — 78 %
Oxygen — 21 %
Carbon dioxide — 0.035 %
[2 marks for all three correct, 1 mark for one or two correct]

d) Deforestation — plants take carbon dioxide out of the atmosphere, so removing trees causes atmospheric carbon dioxide levels to rise *[1 mark]*.
Increased energy consumption — burning fossil fuels releases carbon dioxide (that had been 'locked away') into the atmosphere *[1 mark]*.
Increasing population — *any one of,* more people will be respiring and releasing carbon dioxide / more energy will be used (and carbon dioxide is released when fuels are burnt) / more land will be needed, so trees will have to be cut down, and less carbon dioxide will be removed from the air *[1 mark]*.

---

3 a) i) aluminium *[1 mark]*

ii) *Any one of,* e.g. it is more reactive / more energy is needed / there are many stages to the process / it requires lots of electricity, which is expensive *[1 mark]*.

iii) *Any two of,* e.g. it preserves limited resources / it reduces energy used in the extraction process / it reduces waste / it reduces pollution *[2 marks]*.

b) *Any three of,* e.g. the metal rods should be the same size / the metal rods should be the same distance away from the heat / the same amount of petroleum jelly should be used on each rod / the wooden splints should be the same distance from the end of each rod / the wooden splints should be the same size *[3 marks]*.

---

4 a) i) First three points accurately plotted to +/- half a square *[1 mark]*, second three points accurately plotted to +/- half a square *[1 mark]*.

ii) Smooth curve of best fit drawn (no double lines, no straight lines between points) *[1 mark]*.

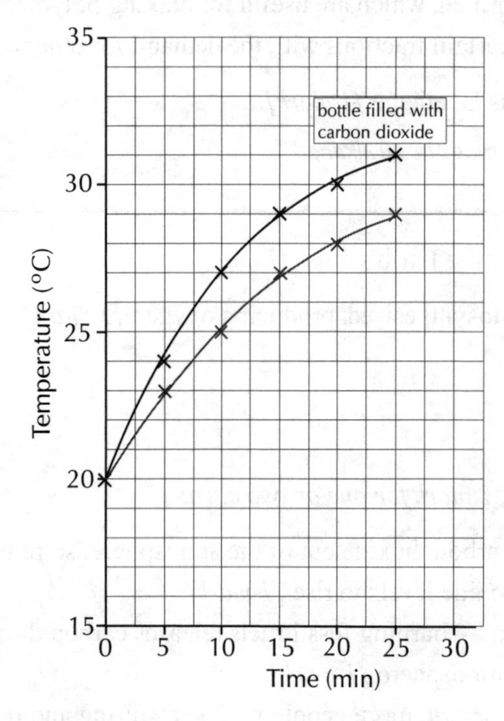

b) The temperature of the carbon dioxide increased more than the temperature of the air *[1 mark]*.

The line showing the temperature of the carbon dioxide reaches a higher point on the graph.

---

5 a) A substance that speeds up a reaction *[1 mark]*.

b) Ethene is produced from crude oil, which is a non-renewable resource *[1 mark]*.

c) ethanol *[1 mark]*

d) E.g. as a fuel / for drinking / as a solvent *[1 mark]*.

---

6 a) carbon/soot and carbon dioxide *[1 mark]*

b) C_6H_{14} + $\mathbf{13O_2}$ = $\mathbf{6CO_2}$ + $\mathbf{7H_2O}$

[1 mark for the correct products, 1 mark for correctly balancing the equation]

c) There was not enough oxygen for complete combustion to take place (soot/carbon is one of the products of incomplete combustion) *[1 mark]*.

Physics Paper 1

Answer			Explanation
1.	**1A**	4	
	1B	1	In solids, the particles are tightly packed in a rigid structure, so their vibrations are easily passed on.
	1C	3	In convection, the material itself moves — solids can't flow like liquids and gases can.
	1D	2	In a vacuum, there are no particles to transfer energy by conduction or convection.
2.	**2A**	1	
	2B	2	
	2C	4	
	2D	3	Gamma radiation can kill living things, including bacteria.
3.	**3A**	1	
	3B	4	2 is also an advantage of using nuclear fuels, but it's not an environmental one.
	3C	2	
	3D	3	63 (fossil fuels) + 16 (nuclear) = 79. So 21% of electricity is generated from other resources.
4.	**4A**	2	Only a small proportion is being 'wasted' (as heat).
	4B	4	It's the thickness of the arrows that matters — number 4 has the thinnest 'light energy' arrow.
	4C	1	$50 \div 200 = 0.25$
	4D	4	Payback time = cost ÷ saving = $375 \div 0.75 = 500$ hours.
5.	**5A**	1	Wavelength = speed ÷ frequency = $(3.0 \times 10^8) \div (225.64 \times 10^6) = 1.33$ m
	5B	4	They travel at the same speed (about 3.0×10^8 m/s).
	5C	3	Frequency = speed ÷ wavelength = $(3 \times 10^8) \div 2 = 1.5 \times 10^8$ Hz
	5D	3	
6.	**6A**	3	90 mins = 1.5 hours. $1.5 \times 1.5 = 2.25$ kWh. $2.25 \times 18 = 40.5$p.
	6B	2	
	6C	4	$1800 \div 5 = £360$ per year. $360 \div 900 \times 100 = 40\%$.
	6D	3	When it isn't windy, the turbine won't generate electricity, so he'll need to buy it from the National Grid.
7.	**7A**	4	Of the four options, a lead lined suit is the only one that will stop beta radiation reaching the body.
	7B	3	The circuit is broken when alpha radiation is absorbed by smoke — stopping it ionising the air.
	7C	2	Alpha is strongly ionising and cannot pass out of the body (not 3 or 4). The half-life should be short.
	7D	3	Alpha would be stopped by any thickness. Gamma would penetrate any thickness.
8.	**8A**	4	The frequency of light is lower than expected (i.e. red-shifted) because they are moving away from Earth.
	8B	1	Galaxies that are further away are moving away from us faster than those that are close.
	8C	2	There are other theories. 3 is not a good reason to accept a theory. Scientific theories cannot be proved.
	8D	2	Light is red-shifted because the galaxies are moving away from us.

Physics Paper 2

1 a) i) *Any one of,* because it doesn't have enough mass / because it is not a heavy-weight star / because it is a middle-weight star *[1 mark].*

The Sun will eventually become a white dwarf.

ii) Because (they are so massive that) their gravitational pull is so strong that even light can't escape. *[1 mark]*

b) i) Mars and Jupiter *[1 mark].*

The asteroid belt lies between the 'rocky' planets and the gas giants.

ii) The strong gravitational pull of Jupiter disrupted the formation of another planet, creating the asteroid belt *[1 mark].*

iii) *Any one of,* e.g. craters on the Earth / layers of unusual elements found in rocks / sudden changes in fossil numbers between adjacent layers in rocks *[1 mark].*

Past asteroid impacts are thought to have changed the conditions on Earth enough to wipe out entire species, causing abrupt changes in the species represented in the fossil record.

---

2 a) The climate is normally very cold and windy but is warmer and calm this year *[1 mark].*

b) Light, shiny surfaces, like ice, reflect more (absorb less) heat radiation than darker surfaces like the surface of the sea *[1 mark].*

c) i) *Any one of,* burning fossil fuels / driving cars / industrial processes / deforestation / cattle rearing *[1 mark].*

ii) *Any one of,* using alternative energy sources / using public transport / planting trees / recycling *[1 mark].*

d) As there becomes less ice, less heat will be reflected and more heat will be absorbed by the Earth *[1 mark]*, so surface temperatures might rise even faster than before *[1 mark].*

---

3 a) i) 2 and 3 *[1 mark]*

ii) 1 because it has the longest wavelength and lowest frequency *[1 mark].*

b) i) visible light and infrared radiation *[1 mark]*

ii) *Any one of, e.g.* they are used as telecommunication cables / they are used in endoscopes for keyhole surgery *[1 mark].*

---

4 a) Wavelength = velocity ÷ frequency = $3 \times 10^8 \div 2.45 \times 10^9 = 0.122$ m (or 12 cm) *[1 mark].*

b) Any one of, communication to and from satellites / mobile phone networks *[1 mark].*

c) Microwaves can be absorbed by water molecules in cells, which heats them up *[1 mark].*

d) The wavelength of visible light is smaller than the holes in the mesh so the light can pass through the holes, allowing the user to see the food *[1 mark]*.

The wavelength of the microwaves (12 cm) is bigger than the small holes in the mesh, therefore the microwaves cannot pass through the holes and harm the user *[1 mark]*.

---

5 a) i) Heat is conducted much more slowly through air than through solid materials like brick *[1 mark]*.

ii) The air is trapped in tiny bubbles in the foam, so convection is reduced *[1 mark]*.

b) They prevent overheating of the house — if the inside is cooler, less heat is lost to the surroundings, so less energy is wasted *[1 mark]*.

c) Air near windows is cooled (by conduction) *[1 mark]* so it contracts / becomes denser and sinks *[1 mark]*.

---

6 a) 214 – 83 = 131 *[1 mark]*

b) The electron is produced in the nucleus when a neutron changes into a proton *[1 mark]*.

c) $^{226}_{88}\text{Ra} \rightarrow {}^{222}_{86}\text{Rn} + {}^{4}_{2}\text{He}$

$^{210}_{82}\text{Pb} \rightarrow {}^{210}_{83}\text{Bi} + {}^{0}_{-1}\text{e}$ *[2 marks]*

d) No, because it emits alpha radiation which cannot penetrate skin, so will stay in the body, and will remain radioactive for a very long time due to its long half-life *[1 mark]*.

ISBN 1 84146 645 X
9 781841 466453

General Certificate of Secondary Education

GCSE Core Science

CHEMISTRY
Paper 1
(Objective Test)

Higher Tier

Centre name				
Centre number				
Candidate number				

Surname
Other names
Candidate signature

Time allowed: 30 minutes.

Instructions to candidates

- Write your name and other details in the spaces provided above.
- Answer all questions on the answer sheet provided.
- Do all rough work on this question paper.
- To record your answers:
  - Use a black ball-point pen
  - Fill in the circle for your answer as shown: 1 ○ 2 ○ 3 ● 4 ○
  - If you want to change your answer, first cross out your original answer as shown: 1 ● 2 ○ 3 ⊗ 4 ○
- You must hand in both your answer sheet and question paper at the end of the test.

Information for candidates

- Marks will not be deducted for incorrect answers.
- In calculations show clearly how you worked out your answers.
- You may use a calculator.
- There are 8 questions in this paper.
- The maximum mark for this paper is 32.

Advice to candidates

- Do not choose more responses than you are asked to.
- Work steadily through the paper.
- Don't spend too long on one question.
- If you have time at the end, go back and check your answers.

For examiner's use

Q	Attempt Nº 1	2	3	Q	Attempt Nº 1	2	3
1				5			
2				6			
3				7			
4				8			
				Total / 32			

[BLANK PAGE]

SECTION ONE

Questions **ONE** and **TWO**.
In these questions, match the letters **A**, **B**, **C** and **D** with the numbers **1** – **4**.
Use **each** answer only **once**.
Mark your choices on the answer sheet.

QUESTION ONE

Match the elements **A**, **B**, **C** and **D** with the numbers **1** – **4** in the following descriptions.

A Carbon

B Sulfur

C Calcium

D Hydrogen

...**1**... is the metallic element present in limestone. C

...**2**... is a non-metal that is used to extract iron from its ore. D

...**3**... chemically combines with **2** to form compounds called hydrocarbons. A

...**4**... dioxide combines with water to form acid rain. B

Turn over for the next question

Turn over➤

QUESTION TWO

Match the words **A**, **B**, **C** and **D** with the numbers **1** – **4** in the passage.

A properties 3

B elements 2

C atom 1

D transition metals 4

Substances made up of only one kind of ...**1**... are called ...**2**... The periodic table contains all known elements — it's organised into groups that contain elements with similar ...**3**... The ...**4**... are found in the middle section of the periodic table.

SECTION TWO

Questions **THREE** to **EIGHT**.
Each of these questions has four parts.
In each part choose only **one** answer.
Mark your choices on the answer sheet.

QUESTION THREE

Read the passage about Alfred Wegener's Theory of Continental Drift.

> Alfred Wegener trained as an astronomer but became interested in fossils while teaching at the University of Marburg. He noticed that identical fossils were found on different continents. His theory was that the continents were once joined together and had then split apart and drifted away from one another.
>
> He published his theory in a German geological journal in 1912 and wrote a book about the subject in 1915. Ten years later his theory was translated into English.
>
> Mapmakers in the 16th century had noticed that the coastlines of continents appeared to fit into one another like a giant 'jigsaw'. Although this observation supported Wegener's theory, very few people initially believed he was correct. One reason was because he suggested that continents could 'plough' through rocks on the ocean bed and people didn't think that this was possible.

3A Which one of the following reasons helps to explain why Wegener's theory was not quickly accepted?

1 People did not believe him because he was German.

2 People thought he had stolen his ideas from 16th century mapmakers.

3 Wegener was not trained as a geologist.

4 There was no evidence to support his theory.

3B Suggest what helped to make Wegener's theory more widely known.

1 His publications were translated into other languages.

2 Scientists began to realise that continents fitted together like a jigsaw.

3 Identical fossils were discovered on different continents.

4 It became possible to take photographs of the Earth from space.

Turn over➤

3C Which part of Wegener's theory do we now know is not correct?

1 Identical fossils exist on different continents.

2 Some continents are moving away from one another.

3 Continents were once joined together.

4 Continents move by 'ploughing' through the ocean beds.

3D What is caused by the sudden movement of tectonic plates?

1 convection currents

2 earthquakes

3 Pangaea

4 radioactive decay

QUESTION FOUR

Metals can be extracted from their ores, which are found in the Earth's crust.

The reactivity of a metal affects how easy it is to extract.
The box shows the reactivity of some elements.

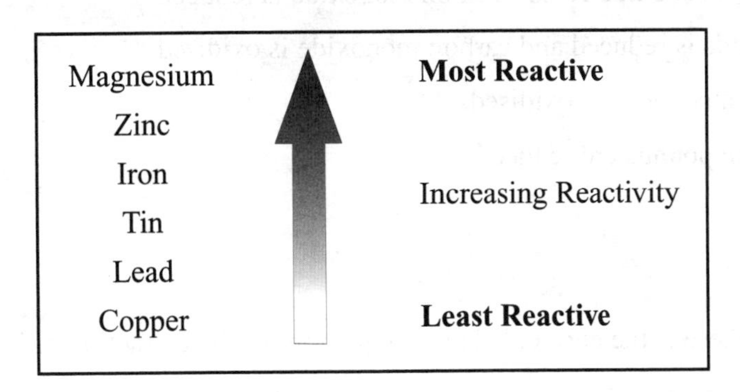

4A Which metal in the list above is most likely to be found uncombined in the Earth's crust?

1 copper
2 magnesium
3 zinc
4 lead

4B Magnesium **cannot** be extracted from its ore by reacting it with carbon because

1 it is an alkali metal
2 it is less reactive than carbon
3 it is more reactive than carbon
4 it is less reactive than iron

Question 4 continues on the next page

4C Iron is extracted from its ore using a blast furnace.
One of the reactions that occurs in the blast furnace is:

iron oxide + carbon monoxide → iron + carbon dioxide

In this reaction

1 iron oxide is oxidised and carbon monoxide is reduced

2 iron oxide is reduced and carbon monoxide is oxidised

3 both compounds are oxidised

4 both compounds are reduced

4D Which of these is the correct balanced equation for the extraction of iron from its ore?

1 $Fe_2O_3 + CO \rightarrow 2Fe + CO_2$

2 $2FeC + O_4 \rightarrow 2Fe + 2CO_2$

3 $Fe_2O_3 + 3CO \rightarrow 2Fe + 3CO_2$

4 $FeSO_4 + CO \rightarrow Fe + CO_2 + SO_4$

QUESTION FIVE

Crude oil is a mixture consisting mainly of hydrocarbons.

The graph shows how the boiling point of a hydrocarbon is related to the number of carbon atoms that it contains.

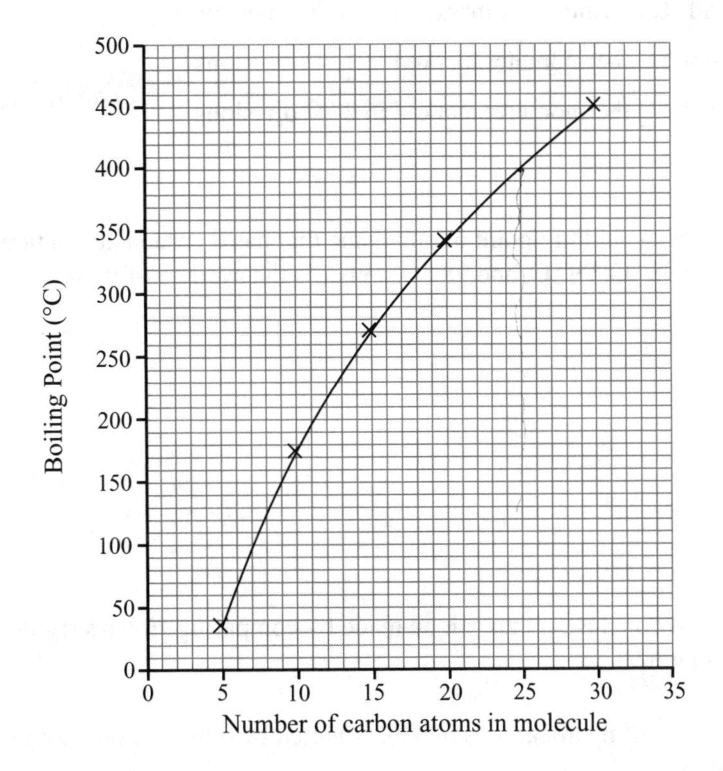

5A From the graph, which is the best estimate of the boiling point of a hydrocarbon with 25 carbon atoms?

1 375 °C

2 450 °C

3 400 °C ✓

4 425 °C

Question 5 continues on the next page

5B Helen wants to measure the boiling points of a range of hydrocarbon molecules as accurately as possible. The hydrocarbons that she will be testing have a range of 1 to 20 carbon atoms. Which piece of equipment would be the most suitable?

1 A –50 to +50 °C thermometer marked in 0.5 °C divisions

2 A 0 to +50 °C thermometer marked in 0.1 °C divisions

3 A 0 to +450 °C thermometer marked in 1 °C divisions

4 A 0 to +600 °C thermometer marked in 2 °C divisions

5C Helen measures the boiling point of an alkane that has 8 carbon atoms three times. Her results are 98 °C, 99 °C, and 98 °C. What is the average of these three readings?

1 97 °C

2 98.3 °C

3 98.5 °C

4 99 °C

5D What is the best conclusion that can be made by comparing Helen's results with those in the graph?

1 Boiling points of hydrocarbons may be affected by other factors, not just the number of carbon atoms.

2 Boiling points of hydrocarbons are not related to the number of carbon atoms.

3 It is difficult to measure the number of carbons in a hydrocarbon accurately.

4 Boiling points cannot be measured accurately for hydrocarbon molecules.

QUESTION SIX

This question is about metals and alloys.

6A An alloy is...

1 a compound of a metal with a non-metal.

2 a pure element.

3 a mixture of a metal with a non-metal or other metals.

4 a mixture of non-metals.

6B Pure iron has fewer uses than steel because...

1 it is too dense.

2 it is too soft.

3 it cannot be shaped easily.

4 it does not conduct electricity well.

6C Alloys are usually harder than pure metals because...

1 the atoms in an alloy are linked by chemical bonds.

2 carbon is a harder material than most metals.

3 transition metals are harder than other metals.

4 the atoms are not all the same size so it is harder for the atoms to slide over each other.

6D Another new technological advance is smart materials.
Which **one** of the following is an example of a smart alloy?

1 nitinol

2 Thinsulate™

3 Teflon®

4 titanium dioxide

Turn over➤

QUESTION SEVEN

The diagram shows four hydrocarbon molecules.

W

```
    H   H   H   H
    |   |   |   |
H—C—C—C—C—H
    |   |   |   |
    H   H   H   H
```

X

```
    H   H   H
    |   |   |      H
H—C—C—C=C
    |   |          H
    H   H
```

Y

```
    H   H   H   H
    |   |   |   |
H—C—C=C—C—H
    |           |
    H           H
```

Z

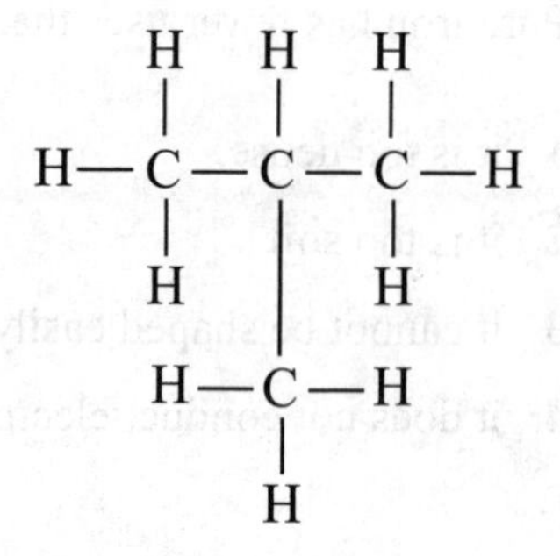

7A What is the chemical formula for molecule **Z**?

1 C_4H_{10}
2 C_4H_8
3 C_3H_{10}
4 C_3H_8

7B Which of these molecules are alkenes?

1 **W, X** and **Y**
2 **W** and **Z**
3 **X** and **Y**
4 all of them

7C Which of these molecules could be turned into a plastic by polymerisation?

1 **X** only

2 **X** and **Y**

3 **Z** only

4 all of them

7D The four hydrocarbons are tested with a chemical. When mixed with **X** and **Y** the chemical changes colour. There is no colour change when it is mixed with **W** or **Z**. What chemical is this?

1 hydrogen

2 steam

3 phosphoric acid

4 bromine water

Turn over for the next question

Turn over➤

QUESTION EIGHT

The table compares the properties of several fuels that can be used in cars.

Fuel	Source	Energy Content (kJ per litre)	Emissions Produced by Burning Fuel
Petrol	Crude Oil	28 000	carbon dioxide, water, carbon monoxide, sulfur dioxide, soot
Ethanol	Fermentation of plants or from ethene	20 000	carbon dioxide, water
Hydrogen	Electrolysis of water or from natural gas	1625*	water
Diesel	Crude Oil	32 500	carbon dioxide, water, carbon monoxide, sulfur dioxide, soot

*when hydrogen is at a safe pressure

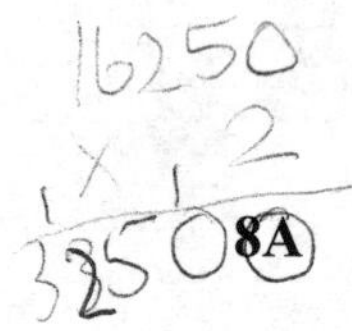

8A Which fuel would contribute the least to global warming when burned?

1 diesel
2 ethanol
3 hydrogen
4 petrol

8B How many litres of hydrogen gas contain the same amount of energy as one litre of diesel?

1 17
2 20
3 0.05
4 12.3

8C Based on the data in the table, which is the best explanation for why hydrogen gas is not more widely used as a fuel for cars?

1 Cars would need very large tanks to store enough hydrogen gas to produce sufficient energy.

2 The raw material needed to make hydrogen is very expensive.

3 Burning hydrogen is dangerous to the environment.

4 It is too difficult to convert car engines to run on hydrogen.

8D Which of the fuels in the table is a pure compound?

1 diesel

2 ethanol

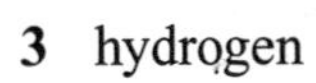

3 hydrogen

4 petrol

END OF TEST